Introduction

Functional Skills Maths Level 1 is aimed at helping you pass this exam with ease.

In the actual test, although the use of calculators is allowed, it is useful to do simple sums with confidence without using calculators. In addition to basic addition, subtraction, multiplication and division you are expected to be familiar with fractions, decimals, percentages, ratios and proportions in everyday context.

Everyday problems often involve being able to estimate as well as being able to work with function machines and formulas like Speed, Distance and Time. Also it is useful to be able to convert from one type of currency to another when you go on holiday! In addition, basic Data Interpretation or Statistics is useful to make sense of data that is presented numerically or visually in a workplace or in newspaper articles. There are also chapters on basic shapes and spaces since you also need to know how to work out lengths around a shape (perimeters), areas and volumes of basic shapes as well as be familiar with plans and drawing nets!

Whichever Exam Board you are working for you will find this book useful for Functional Mathematics up to level 1 and prepare you for Functional Mathematics Level 2.

These tests may be taken online and in the actual tests you can either pass or fail.

Although you are allowed to use calculators it is very easy to press the wrong key so it is good to check your answers by doing it again or step by step. If possible at least estimate the value of the answer. Then you will know whether the answer is sensible or not. This will help you to approach basic arithmetical problems with more confidence.

One thing to remember is that there is often more than one way of working out a given problem. It does not matter which method you use, so long as you feel comfortable with it.

About the Author

The author of this book has experience in both consultancy work and teaching.

The author's initial book 'Speed Mathematics Using the Vedic System' has a significant following and has been translated into Japanese and Chinese as well as German. In addition, his book 'Pass the QTS Numeracy Test with ease' is very popular with teacher trainees. Besides being a specialist mathematics teacher the author also has a degree in psychology. Besides working in consultancy he also managed the QTS numeracy tests for teacher trainees at OCR in conjunction with the teaching agency. Subsequently he has tutored and taught mathematics and statistics in schools as well as in adult education. He is also the author of 12 other maths books.

He hopes that this book 'Functional Maths Level 1' will help those aspiring to pass this test which ever exam board you are studying for.

Chapter 1   Arithmetic part I

Numbers:

Numbers as you recall are made of digits (0, 1, 2, 3, 4, 5, 6, 7, 8 and 9)

Example1: Write the number two hundred and fifty three

This is simply 253

Hundreds   Tens   Units

Example 2: Write three thousand, four hundred and sixty five as a number.

Method: This is written as 3465. (The first digit '3' represents three thousand, the second digit '4' represents four hundred, '6' represents six tens and '5' represents five units.

Example 3: Write in words the number 56342

This as you have probably guessed is **fifty six thousand, three hundred and forty two.**

Example 4: Write the numbers 456, 23, 459, 12, 1031, 98, 21 from smallest to biggest.

Method: Starting with the smallest number first the numbers can be written as **12, 21, 23, 98, 456, 459, 1031 (you can see the numbers get bigger as we go along)**

Addition, Subtraction, Multiplication and Division

Remember you can always use a calculator in the exam.

I am sure you know that 2 + 3 = 5

Hopefully you don't need to use a calculator for this.

But what about 9 + 25?

Yes you can use a calculator or you can use the method below:

Method: (adjusting or compensating method)

9 is the same as 10 – 1 so 9 + 25 = 10 – 1 + 25 = 10 + 25 – 1 = 35 -1 = 34!

Or as I said you can always use a calculator.

Let us look at some real life examples.

<u>**Example 1:**</u>

Joanna goes to buy some groceries for her dinner after a hard day's work. She decides to buy all the three items listed below. How much does she pay altogether?

$\frac{1}{2}$ **Roast Chicken for £3**

Tomato soup for £2

A loaf of whole meal bread for £1

<u>**Method:**</u> Total spent is £3 + £2 + £1 =£6

Simply add up the cost of all the items. So the total Joanna spends = £6 **(don't forget to put the £ sign in front of the 6!**

<u>**Example 2:**</u>

John is doing some decorating. He buys a tin of glossy white paint and a good quality brush. The prices are given below. How much does he spend in total?

Tin of glossy white paint: £9

Paint brush: £8.50

Method: The total John spends is 9 + 8.50 = £17.50

(If you are doing this by a calculator make sure to put the decimal point after '8' in 8.50)

Subtraction

Example1:

Work out: 241 – 28

Method: Using a calculator we find that 241 – 28 = 213

Example 2: Miranda has saved £60. She buys a pair of shoes for £29. How much does she have left? **As usual use a calculator if you find it easier.**

Method: 60 – 29 = £31

You can check this by adding £29 to £31 to get the original amount which is £60.

Practice Question 1 - on Numbers, Adding and Subtracting

(You can use a calculator from number 4 onwards if you wish)

(1) Write the number 3045 in words

(2) Write five hundred and two as a number

(3) Write thirty two thousand, three hundred and fifty as a number

(4) Jeremy who is self-employed and works in the building trade has 3 days of work one particular week. He earns £90, £85 and £95 for the days respectively. How much in total does he earn?

(5) Jemma is planning a short holiday. She has saved £490. The holiday costs her £370. How much does she have left?

(6) Ahmed orders a book from Amazon for £6.50. He has to pay £2.80 for postage and packing. How much does he pay in total?

(7) Miranda goes out with her friend for lunch and decides to pay for it. The bill comes to £18.75. Miranda gives a £20 note. How much change does she get?

(8) Fatima goes shopping. She buys a handbag for £19.80, a blouse for £12.50 and a scarf for £11.75. How much does she spend altogether?

(9) John has £40 and Joseph has £75. How much more does Joseph have compared to John.

(10)   Joanna books a holiday for £375. However as she books online she gets a discount of £30. How much does Joanna pay for the holiday?

(11)   Peter manages to save £235 over 4 months. He decides to buy a computer tablet for £198. How much savings does Peter have after his purchase?

(12)   In the table below work out the total in column B

A	B	C
£13.75	£102	£122
£12.00	£220	£131
£101.50	£150.50	£95.50

Answers to Practice Questions 1 - on Numbers, Adding and Subtracting

(1) Three thousand and forty five

(2) 502

(3) 32,350

(4) £270

(5) £120

(6) £9.30

(7) £1.25

(8) £44.05

(9) £35

(10) £345

(11) £37

(12) £472.50

Multiplication of whole numbers by 10, 100, 1000.

Multiplying a number by 10 simply means it becomes 10 times bigger.

It is useful to know that when you multiply a **whole number** by **10** you just add one zero to the number at the end. When you multiply a **whole number** by a **100** you add two zeros at the end and so on. Some examples below will help you understand this more clearly. Multiplying by 10 means the answer is 10 times bigger, similarly multiplying by 100 means the answer is 100 times bigger. (**Note: This method only works for whole numbers!**)

Examples:

(1) 45 × 10 = 450 (**add 1 zero to 45**)

(2) 67 × 100= 6700 (**add 2 zeros to 67**)

(3) 65 × 1000= 65000 (**add 3 zeros to 65**)

Example: A container for eggs has 6 eggs in it. I buy 10 such containers. How many eggs do I have altogether?

Method: Add '0' to 6 to get 60 eggs altogether. You can of course use a calculator by multiplying 6 by 10!

Decimal points.

Look at the number line below.

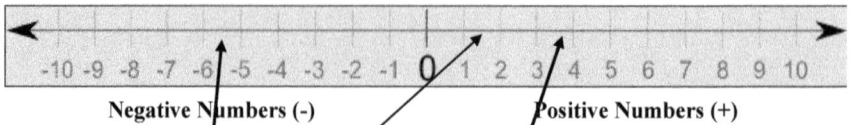

Negative Numbers (-)   Positive Numbers (+)

(1) You can see that 1.5 is half way between 1 and 2.

(2) Similarly 3.6 is more than three but less than 4.

(3) If you have a number like -5.34 it might be difficult to show but you know it's roughly here.

When adding, subtracting, dividing or multiplying don't forget to put the decimal point in the calculator.

Easy rules for numbers with decimals when multiplying by 10

When multiplying by 10, 100, 1000 move the decimal place the appropriate number of places to the right. **You can always use a calculator if you prefer!**

(1) 67.5 × 10 = 675 (the decimal point is moved 1 place to the right to give us 675.0 which is the same as 675)

(2) 67.5 × 100 = 6750 (this time move the decimal point two places to the right to give 6750.0 which is the same as 6750)

(3) 6.87 ×1000 = 6870 (in this case move the decimal point three places to the right to give the required answer.)

Division

Now consider examples involving division by 10, 100 and 1000 and other powers of ten.

(1) 450 ÷ 10 = 45 (You simply remove one zero from the number)

(2) 5600 ÷ 100 = 56 (This time you remove two zeros from the number)

(3) 45 ÷ 100 = 0.45 (No zeros to remove – so this time move the decimal point two places to the left to give us 0.45)

(4) 345.78 ÷ 100 = 3.4578 (Again simply move the decimal point 2 places to the left to give the answer)

(5) 456.78 ÷ 1000 = 0.45678 (Move the decimal point 3 places to the left as shown)

Questions involving powers of 10 (use a calculator or the 'easy method' whichever you prefer)

(1) Divide 27000 by 10

(2) What is 7887 multiplied by 100?

(3) What is 67 divided by 100?

The answers are:

(1) 2700   (2) 788700 (3) 0.67

Some other Multiplication Examples

Example 1: work out 14×13

Using a calculator we find that 14×13 = 182

Example 2: In a certain company 54 insurance agents manage to sell 14 insurance policies each in a particular month. How many insurance policies did these agents sell altogether in that month?

Method: Using a calculator we find the total insurance policies sold were 54 × 14 = 756

Multiplying by 5 quickly (Just in case you are interested to know)

Multiply the number by 10 and halve the answer.

Example 1: 5 × 4 = half of 10 × 4 = half of 40 = 20

Example 2: 5 × 16 = half of 10 × 16 = half of 160 =80

Example 3: 5 × 23 = half of 10 × 23 = half of 230 = 115

Square numbers

10^2 (10 squared) means 10×10 = 100

Similarly 5^2 = 5×5 = 25, 7^2 = 7×7 =49, 15×15 = 225 and so on.

Example 1: Square the number 25

Method: We simply multiply 25×25 which equals 625

Example 2: Find the value of 18^2

Method: 18^2 means 18×18 = 324

Summary: To square a number you simply multiply the number by itself.

The Order of Arithmetical Operations

Remembering the order in which you do arithmetical operations is important.

(1) Always work out the **Bracket(s)** first
(2) Now **Multiply** and **Divide**
(3) Finally do the **Addition** and **Subtraction**.

Example 1: 4 + 13(7 − 2) this means add 4 to 13× (7 − 2)

Do the brackets first so (7 − 2) = 5, then multiply 5 by 13 to get 65 and finally add 4 to get 69

Example 2: Work out 2 + 8×3

Do the multiplication before the addition

So 8×3 =24 and 2 + 24 = 26

Example 3: Work out (2×15) + (4×8)

Do the bits in the brackets first: So we have (2×15) = 30 and (4×8) = 32

Now add 30 to 32 to get 62.

> <u>Summary</u>: When working out sums involving mixed operations (e.g. +, - , x and ÷) you need to work out the steps in stages using the above rules: So to work out 8 +25 ×12. Do the multiplication first, 25×12 =300, write down 300 then add 8 to get the answer 308. (When using a calculator be careful that you follow the above rules otherwise you could get the wrong answer!)

Division

Dividing a number by 2 is a very useful skill, since if you can divide by 2, you can by halving it again divide by 4.

Dividing by 2 and 4

Simply halve the number to divide by 2

Halving the number again is the same as dividing by 4

Example 1: 28 ÷ 2 =14 (half of 28)

Example 2: $48 \div 4 = 24 \div 2 = 12$ (check using a calculator)

Example 3: $64 \div 4 = 32 \div 2 = 16$ (check using a calculator)

Dividing by 5

An easy way to do this is to multiply the number by 2 and divide by 10.

Example 1: $120 \div 5 = (120 \times 2) \div 10 = 240 \div 10 = 24$

Example 2: $127 \div 5 = (127 \times 2) \div 10 = 254 \div 10 = 25.4$

Dividing by other numbers: It is best to use a calculator.

Question involving division

Example 1: In one particular week in a restaurant a bonus of £67.50 is divided amongst three waiters. How much does each waiter get in that week?

Method: £67.5 ÷ 3 = £22.50 per waiter

(In the exam for questions like the one above it is best to use a calculator)

Chapter 2   Positive and Negative Numbers

Positive and Negative Numbers using the number line

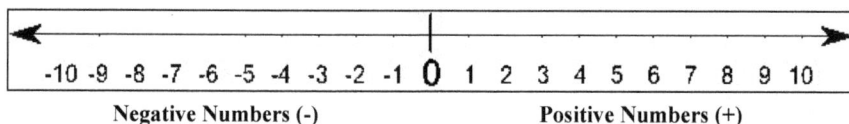

```
◄─────────────────────────┼──────────────────────►
   -10 -9 -8 -7 -6 -5 -4 -3 -2 -1  0  1  2  3  4  5  6  7  8  9  10
       Negative Numbers (-)                    Positive Numbers (+)
```

Numbers on the left decrease in value and numbers on the right increase in value.

Examples:

    (1) **3** is smaller than **5**

    (2) **-5** is smaller than **0**

    (3) **-9** is smaller than **–4**

Numbers on the right are bigger than numbers on the left.

Examples:

    (4) **9** is larger than **4**

    (5) **1** is larger than **–1**

    (6) **5** is larger than **–8**

You don't need to always use the number line. You can also use the method below:

When adding and subtracting positive and negative numbers it is worth knowing the following:

When you add two minus numbers you get a bigger minus number.

Example 1: $-4 - 6 = -10$

When you add a plus number and a minus number you get the sign corresponding to the bigger number as shown below:

Example 2: $+6 - 9 = -3$, whereas, $-6 + 9 = 3$

Practice Questions 2

(1) I can put 12 marbles in a bag. How many marbles can I put in 7 similar bags?

(2) A teacher can put 15 text books in a box. How many text books can a teacher put 14 boxes of the same size?

(3) John earns £11 per hour doing general maintenance work. The amount of work fluctuates. One good week he works for 43 hours. How much does he earn in this particular week?

(4) What is 300 divided by 6?

(5) What is 4000 divided by 100?

(6) How many sevens are there in 147?

(7) John has 400 kilograms of cement which he wants to divide amongst 10 workers. How much does each worker get?

(8) Elizabeth buys some flowers. Each bunch of mixed flowers costs £3.50. Elizabeth buys 6 bunches of mixed flowers. How much does she pay in total?

(9) Which is bigger -7 or 8?

(10) The maximum temperature in London one December was 13 degrees centigrade and the minimum temperature was -3 degrees centigrade. What was the difference between the highest and lowest temperature?

(11) Arrange the numbers -4, -5, 0, -9, 1, 8, -6 from smallest to biggest.

(12) Multiply (a) 6 ×7 and (b) 13×12

Answers to Practice 2

(1) 84 marbles in total

(2) 210

(3) £473

(4) 50

(5) 40

(6) 21

(7) 40 kgm

(8) £21

(9) 8

(10)   16°C

(11)   -9, -6, -5, -4, 0, 1, 8

(12)   (a) 42   (b)156

Chapter 3   Arithmetic Part 3

Time Based Questions

For converting time from 12 hour clock to 24 hour clock - see examples below

12 –Hour Clock	24 –Hour Clock
8.45 am	08:45
11.30 am	11:30
12.20pm	12:20
2.35 pm	14: 35 (after 12pm add the appropriate minutes and hours to 12 hours, in this case 2hrs 35mins +12hrs = 14:35)
8.45 pm	20:45 (8hrs 45mins + 12hrs = 20:45)
11.47pm	23:47 (11hrs 47mins +12hrs = 23:47)

The Convention is that if the time is in 24-hr clock there is no need to put hours after the time.

Also remember: 2.5 hours = 2 hours 30minutes (0.5 hours = half of 60 minutes)

2.25 hours = Two and a quarter hours = 2hrs 15 minutes

2.4 hours = 2 hours 24 minutes (0.4 hours = 0.4X60 = 24 minutes)

2.1 hours = 2 hours 6 minutes (0.1hours = 0.1 X 60 = 6 minutes)

Remember there are 60 minutes in one hour so 0.4 hours = 0.4×60 =24 minutes.

For other time based questions e.g. years, months, days, hours, minutes or seconds remember the appropriate units.

Example 1: At a company new candidates are mentored once a week for 12 minutes each. There are 15 candidates who are being mentored. The session starts at 11.30am. When does it finish? Give your answer using the 24 hour clock

Method: Clearly we need to first work out the total time it takes for all the candidates. Total time for 15 candidates is 15 × 12 = 180 minutes = 3 hours. So the mentoring session ends 3hrs after 11.30am – this means it ends at 2.30pm. However using the 24 hour clock the times it ends is 14:30

Example 2: Peter completes a lap in 2.3 minutes. How many minutes and seconds is this?

Convert 0.3 minutes into seconds. Since one whole minute = 60 seconds, then 0.3 minutes = 0.3X60 = 18 seconds. Hence Peter completes the lap in 2 minutes and 18 seconds.

(Note that 0.3 X 60 is the same as 3 X 6, hence this is equivalent to 18)

General Multiplication questions

Example 1 There are 4 medium size boxes containing 18 black jumpers each and 3 bigger boxes containing 23 black jumpers each. How many black jumpers are there altogether?

Method: 4 boxes of 18 each means there are 4 X 18 = 72 black jumpers

Similarly, 3 boxes of 23 each means, 3 X 23 =69 black jumpers

Adding all the black jumpers we get 72 + 69 = 141

There are a total of 141 black jumpers altogether.

Example 2

I buy 5 books for £3.97 each. How much change do I get from a £20 note?

Method: 5×3.97 = £19.85. Now subtract £19.85 from £20 and you can see that I get 15p change from my £20 note.

(Since 20 – 19.85 = 15)

HARROW COLLEGE
Learning Centre

Rounding numbers and estimating

We will start simply with rounding numbers to the nearest 10 and 100

Consider the number 271

Rounded to the nearest 10 this number is 270

Rounded to the nearest 100 this number is 300

(The principle is that if the right hand digit is lower than 5 you drop this number and replace it by 0. Conversely if the number is 5 or more drop that digit and add 1 to the left)

Try a few more:

5382 to the nearest 10 is 5380

5382 to the nearest hundred is 5400

5382 to the nearest 1000 is 5000

This rule can also be applied to decimal numbers:

3.7653 rounded to the nearest thousandth is 3.765

3.7653 rounded to the nearest hundredth is 3.77

3.7653 rounded to the nearest tenth is 3.8

3.7653 rounded to the nearest unit is 4

Tip: Remember to use common sense when rounding in real life situations – see examples below:

Example 1: A book store wants to keep 120 books in the same size boxes. They can fit 22 books in a box. How many boxes will they need?

Method: Number of boxes required will be 120÷22= 5.5 (to one decimal place). But clearly, they cannot have 5.5 boxes. **So they need to have 6 boxes**

Example 2: Mary can keep 22 pieces of cakes in packs of 5 (**that is 5 pieces per pack**). How many packs will she need?

Method: Number of packs required is 22÷5 = 4.4. But obviously she can't have 4.4 packs so she must have **5 packs to fit all her 22 pieces of cakes.**

Estimating calculations quickly

Example 1: Work out (2.2 × 7.12)/4.12

We can quickly estimate that this is roughly equal to (2 X 7)/4 =14/4 which is around 3.5 or 4 rounded to the nearest unit. The actual answer is: 3.8 (to 1 decimal place)

Example 2: Work out 38 × 2.9 × 0.53

We can approximate 38 to be 40 to the nearest ten

We can approximate 2.9 o 3 to the nearest unit

We can approximate 0.53 to 0.5 to the nearest tenth

So the magnitude of the answer is 40 × 3 × 0.5

This is 120 × 0.5 =60 (approximately)

Practice Questions 3

(1) Work out 123×24

(2) What is 159 ÷3?

(3) Round the number 3.167 to two decimal places

(4) There are 5 boxes containing 19 pairs of shoes each and 3 bigger boxes containing 27 pairs of shoes each. How many pairs of shoes are there altogether?

(5) I buy 5 plates for £1.98 each. How much change do I get from a £10 note?

(6) A teacher wants to keep 140 books in the same size boxes. She can fit 18 books in a box. How many boxes will she need?

(7) In a workplace a new apprentice is trained once a week for 25 minutes. The session starts at 10.30am. When does it finish? Give your answer using the 24 hour clock.

(8) John completes his morning jog in 12.4 minutes. How many minutes and seconds is this?

(9) Round the number 238 to the nearest 10.

(10) Anne manages to save £14.50 per week. How much will she save in one year?

Answers to Practice Questions 3

(1) 2952

(2) 53

(3) 3.17

(4) 176

(5) 10p

(6) 8 boxes (Since you cannot have 7.78 boxes!)

(7) 10:55

(8) 12 minutes 24 seconds

(9) 240

(10) £754

Chapter 4　　　　　　Arithmetic Part 4

Fractions

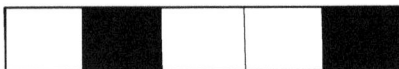

A fraction **is a part of a whole**. So if there are 5 parts altogether and 2 parts are shaded then this can be expressed as 2 out of 5 or $\frac{2}{5}$

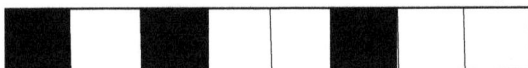

Similarly if 3 parts out of 8 are shaded then this can be written as $\frac{3}{8}$.**The top number is the number of parts you are interested in** and **the bottom number is the total number of parts.**

Example 1: I cut a cake into 9 slices and eat two of them. What fraction of the cake have I eaten?

Method: There are 9 parts in the whole cake. I eat 2 parts out of 9. This means the fraction is $\frac{2}{9}$

I am sure most of you are aware that 'half' can be written as $\frac{1}{2}$

<u>**Try and remember the following fractions**</u>

$\frac{1}{2}$ **(Half)**　　$\frac{1}{4}$ **(a quarter)**　　$\frac{3}{4}$ **(three quarters)**　　　　$\frac{1}{10}$ **(a tenth)**

$\frac{1}{3}$ **(a third)**　　$\frac{1}{5}$ **(a fifth)**

Mixed Fractions: This is simply when you combine a whole number with a fraction, for example $2\frac{1}{2}$ **(Two and a half) or** $1\frac{1}{3}$ **(One and a third)**

Summary:

It is useful to remember some of the following fractions if you can.

$\frac{1}{2}$ (Half), $\frac{1}{4}$ (Quarter), $\frac{3}{4}$ (Three Quarters), $\frac{1}{5}$ (One Fifth), $\frac{1}{10}$ (One Tenth)

Mixed Number: $1\frac{1}{2}$ (One and a half), $2\frac{3}{4}$ (Two and three quarters), etc.

Questions on fractions

Example 1: I cut an orange into 4 equal parts. I eat one part. Express this as fraction of the whole orange.

Method: There are 4 parts in total so the fraction of the orange that I eat is $\frac{1}{4}$

Example 2: A marketing department has 7 people, four of these are women.

(1) What fraction of the marketing department consists of women?

(2) What fraction consists of men?

Method: (1) There are 7 people in total this marketing department and 4 are women, this means the fraction that consists of women is $\frac{4}{7}$

(2) If 4 out of 7 are women this means 3 out of 7 are men. So the corresponding fraction of men is $\frac{3}{7}$

Questions involving fractions

Example 1: Finding a fraction of an amount

(1) Find $\frac{3}{4}$ of £600.

To do this, we first write the above question as $\frac{3}{4}\times$ £600. (**Put '3' in a calculator divide by 4 and multiply the answer by 600. The answer is 450. Since the units are £, the actual answer is £450**

Example (2) Find $\frac{1}{5}$ **of 400 dollars.**

As before to do this, we first write the above question as $\frac{1}{5}\times$ 400. (**Put '1' in a calculator divide by 5 and multiply the answer by 400. The answer is 80. Since the units are dollars, the actual answer is $80 (80 dollars)**

Practice Questions 4:

(1) Write down the following fractions, the first two are done for you.

(a) One quarter can be written as $\frac{1}{4}$, (b) Three sevenths can be written

as $\frac{3}{7}$, (c) Five elevenths can be written as ……. (d) Three tenths can

be written as …….. (e) Four nineteenths can be written as……….
(f) Three fifths can be written as …………

(2) Find $\frac{1}{2}$ of £600

(3) Lucy cuts a whole cake into twelve parts and 3 people eat seven parts
of the cake between them. What fraction of the cake is eaten in total?

(4) Joseph is keen to cut the overgrown lawn in his garden before it starts
raining. He manages to complete $\frac{4}{5}$ of it. Write this fraction in words.

(5) Work out $\frac{4}{5}$ of £600

(6) Mary gives pocket money to her two children. She gives $\frac{3}{5}$ of £5 to the
older child and remainder to the younger child. (a) How much does the
older child get? (b) How much does the younger child get?

(7) Write down three quarters as a fraction.

(8) Fiona and Jacob have £35 between them. Fiona has £22. What fraction
of the total does Fiona have?

(9) A sweater at Marks & Spencer's is on offer at half price. Its original
price was £26. How much will I have to pay if I buy it?

(10) In an engineering factory workforce of 375 people there are 12
women and the rest are men. What fraction of the workforce consists of
women?

(11) Sarah buys a dress on ebay for £20. A week later she manages to sell it for $\frac{3}{4}$ of the price she paid. (a) How much does she sell it for? (b) How much loss does she make?

(12) Macy sells her old car at half the original price. She paid £2400 originally. How much loss did she make?

(13) A sweater is priced at £36. But there is a clearance sale and I can buy it at a quarter of the price. How much do I pay for the sweater?

Answers to Practice Questions 4

(1) (c) $\frac{5}{11}$, (d) $\frac{3}{10}$ (e) $\frac{4}{19}$ (f) $\frac{3}{5}$

(2) £300

(3) $\frac{7}{12}$

(4) Four Fifths

(5) £480

(6) Older child gets £3 and the younger one £2

(7) $\frac{3}{4}$

(8) $\frac{22}{35}$

(9) £13

(10) $\frac{12}{375}$

(11) (a) £15 (b) £5

(12) £1,200

(13) £9

Chapter 5   Arithmetic Part 5   More Fractions

Simplifying fractions

Reducing a fraction to its simplest form.

You can do this two ways (1) **you can divide the top number by the bottom number and the answer will be expressed as a decimal.**

Example1: Reduce $\frac{200}{500}$ to its simplest decimal form. If you convert this fraction into a decimal using a calculator you get 0.4. (Simply put 200 into a calculator and divide by 500)

Or you can divide the top and the bottom number the same number. Both the top and bottom number have to be divisible by it!

Example 2

Reduce $\frac{70}{120}$ to its simplest form. If you divide top and bottom by 10 you get $\frac{7}{12}$

Another example:

Example 3: Simplify $\frac{2}{4}$

Method: You can probably see that $\frac{2}{4}$ is the same as $\frac{1}{2}$. This can be worked out as follows: Divide top and bottom of the fraction $\frac{2}{4}$ by 2. This means the top number becomes '1' and the bottom number becomes '2'. So a simpler way to write $\frac{2}{4}$ is $\frac{1}{2}$

Example 1:

60 people apply for a certain job vacancy. 12 people are short-listed for an interview. What proportion of people that are short listed for the interview? Give your answer as a decimal.

Total number of people applying for this job = 60. Since 12 people are shortlisted. Hence the proportion that are shortlisted = 12/60. If you divide the top number by the bottom number (using a calculator) you get 0.2. The answer as a decimal is thus 0.2

Example 2: Simplify $\frac{6}{9}$ in its lowest form expressing the answer as a fraction.

Method: Take the fraction $\frac{6}{9}$ and divide top and bottom by 3 to get $\frac{2}{3}$

Notice: This time we can't divide top and bottom by 2 since '2' does not divide exactly into 9. So we choose 3 since '3' goes into both the top number and the bottom number exactly.

Practice Questions 5

(1) Work out half of £600

(2) Karen bakes small ginger cakes for a party. For each of these cakes she needs 2 ounces of sugar. If she bakes 20 of these cakes how many ounces of sugar does need?

(3) Richard gets 25 days of holidays per year. He has already used $\frac{3}{5}$ of his holidays. How many days has he got left?

(4) Helen has used up 5 gigabytes of her 8 gigabytes memory on her smart phone. What fraction of the memory does she have left?

(5) Elizabeth manages to book a £210 holiday online with $\frac{1}{3}$ of the price off. What price does she pay for the holiday?

(6) Ahmed has worked overtime for 16 hours this month. His normal pay is £8 per hour and if he works overtime he gets $1\frac{1}{2}$ times as much. How much overtime pay does Ahmed get this month?

(7) Which is bigger $\frac{1}{2}$ or $\frac{1}{3}$?

Answers to Practice Questions 5

(1) £300

(2) 40 ounces

(3) 10 days left

(4) $\frac{3}{8}$

(5) £140

(6) £192

(7) $\frac{1}{2}$ is bigger

Chapter 6   Proportions and ratios

Although proportion and ratio are related they are not the same thing – see example below for clarification.

Example: In a class there are 13 girls and 10 boys. The **ratio of girls to boys is** 13:10, and the **proportion of girls in the class** is 13 out of 23 or $\frac{13}{23}$ (Since the total number of pupils is 23, the bottom number is 23)

Questions based on proportions and ratios

Example 1

In a class of 27 pupils, 9 go home for lunch. What is the proportion of pupils that go home for lunch?

Since 9 out of 27 pupils go home, this means the proportion of pupils that go home for lunch is simply $\frac{9}{27}$.

Example 2:

£100 is divided in the ratio 1: 4 how much is the bigger part?

The total number of parts that £100 is divided into is 5 (to find the number of parts simply add the numbers in the ratio, which in this case is 1 and 4)

Clearly, 1 part equals £20 (100 divided by 5), so 4 parts is equal to£80 (Since 4×20 =80). So £80 is the bigger part.

Example 3:

£1500 is divided in the ratio of 3 : 5 : 7

Find out how much the smallest part is worth?

Clearly £1500 is divided into a total of 15 Parts (add up the ratio parts 3 : 5 : 7

So each part is worth £100 (£1500 divided by 15)

So 3 parts (this is the smallest part) equals 3 ×100 = £300

Example 4:

As we have seen, sometimes ratios are expressed in ways, which may not be the simplest form. Consider 5:10

(a) You can re-write 5:10 as 1:2 (divide both sides of 5:10 by 5)

(b) 4: 10 can be re-written as 2: 5 (divide both sides of 4:10 by 2)

(c) 15 : 36 simplifies to 5 : 12 (divide both sides of 15:36 by 3)

Example 6: A team of 10 people can deliver 6000 leaflets in a residential estate in 3 hours. How long does it take 6 people to deliver these leaflets?

Method: 1 person will take 10 times as long or 3X10 = 30hours

This means 6 people will take 30÷6 =5 hours

Conversions

Conversions are often useful in changing currencies for example from pounds to dollars or euros and vice-versa. It is also useful to convert distances from miles to kilometres or weights from kilograms to pounds and so on.

Basically a conversion involves changing information from one unit of measurement to another. Consider some examples below:

Question based on conversions

Example 1:

I go to France with £150 and convert this into Euros at 1.2 Euros to a pound.

(1) How many Euros do I get? **(2)** I am left with 39 Euros when I get back home. The exchange rate remains the same. How many pounds do I get back?

Method: (1) Since 1 pound = 1.2 Euros, I get 150 X 1.2 =180 Euros in total.

(2) When I get back I change 39 Euros back into pounds. This time I need to divide 39 by 1.2

So 39÷1.2 =32.5. This means I get back £32.50

Example 2

The formula for changing kilometres to miles is given by:

M = $\frac{5}{8}$X K. Use this formula to convert 68 kilometres to miles

Method: substitute **K** with 68 and multiply by $\frac{5}{8}$

This means M = $\frac{5}{8}$X 68. Using a calculator this comes to 42.5 miles

It is worth reviewing some common Metric and Imperial Measures as shown below

Units, Weight and Capacity

Metric Measures

1000 Millilitres (ml) =1 Litre(l)

100 Centilitres (cl) =1 Litre (l)

10ml =1 cl

1 Centimetre (cm) =10 Millimetres (mm)

1 Metre (m) = 100 cm

1 Kilometre (km) =1000 m

1 Kilogram (kg) =1000 grams (g)

Imperial Measurements

1 foot =12 inches

1 yard =3 feet

1 pound = 16 ounces

1 stone =14 pounds (lb)

1 gallon = 8 pints

1 inch = 2.54 cm (approximately)

Question on conversions

Example 1: How many grams are there in 2.5kg?

Method: Each Kg = 1000g. So 2.5kg = 2.5×1000 g = 2500g

Example 2: How many cm are there in 0.5km?

Method: Since, 1metre = 100 cm and 1km = 1000m, this means 0.5km = 500m = 500×100 cm = 50000 cm

Example 3: Robert is going to the USA. He changes £120 into US dollars. The exchange rate after Brexit is $1.2 to one pound sterling. How many dollars does Robert get?

Method: Since we know that one £ sterling = $1.2 this means £120 = $120×1.2 = $144

Example 4: A ramblers' group, go on a walking tour whilst in the South of France. They walk from Perpignan to Canet Plage which is approximately 11 km away. After a lunch break and some time on the beach, they walk back to Perpignan. How many miles in total do they walk on that day? (You are given that 8 km is approximately equal to 5 miles.) Give your answer as a decimal.

Method: Total distance walked = 22Km (11 + 11). To convert this into miles we have to multiply 22 by 5 and then divide by 8 (Since 8 km = 5 miles)

That is 22X $\frac{5}{8}$ =13.75 miles (divide 5 by 8 and multiply the answer by 22)

Practice Questions 6

(1) Mary invites 28 people to her birthday party. 23 of them are girls. What is (a) the proportion of girls in the party? (b) What is the proportion of boys?

(2) Helen cycles to work 3 days out of 5 during a working week. What is the proportion of days she cycles to work?

(3) Ahmed has sandwiches 5 times a week for his lunch. For 2 days he has curry. What is the proportion of times he has sandwiches for lunch?

(4) The ratio of sugar to fibre in an orange is 1:7. Assuming an orange weighs 80 grams how many grams of sugar does it contain?

(5) £500 is split between John and James in the ratio of 2:3. How much does James get?

(6) £1200 is divided between 3 people in the ratio of 1:2:3. What is the largest amount that a person gets?

(7) In 2014 in a particular school 79 people take French out of a total of 690 pupils. What is the proportion of pupils who take French in this school?

Answers to Practice Questions 6

(1) (a) The proportion of girls are $\frac{23}{28}$

(b) The proportion of boys are $\frac{5}{28}$

(2) $\frac{3}{5}$

(3) $\frac{5}{7}$

(4) 10gms

(5) £300

(6) £600

(7) $\frac{79}{690}$

Chapter 7   Percentages

Percent simply means out of 100. (Per Cent = Per 100)

The symbol used for per cent is %

So 10% means 10 out of 100 or $\frac{10}{100}$

So 12% means 12 out of 100 or $\frac{12}{100}$

<u>To work out percentages can be very useful.</u>

Example 1. Find 30% of £40

We can work this out by writing $30\% = \frac{30}{100}$

So 30% of £40 can be written as $\frac{30}{100} \times 40$ (Remember 'of' in maths can be written as times '×')

Now to work out $\frac{30}{100} \times 40$ we simply use the calculator by dividing 30 by 100 and then multiplying the answer by 40. That is $30 \div 100 \times 40 = £12$

Example 2: Find 25% of £600

Using the method above we get $\frac{25}{100} \times 600 = 25 \div 100 \times 600 = £150$ (As always you can use a calculator to work this out)

Example 3: I buy a pair of shoes whose original price is £50. However in a sale there is a discount of 30%. How much do I pay for the pair of shoes?

Method: First work out 30% of £50 $= \frac{30}{100} \times 50$, using a calculator we find $30 \div 100 \times 50 = £15$

We know that the discount is £15. So we pay £50 - £15 = £35 for this particular pair of shoes.

Practice Questions 7

(1) Find 20% of £500

(2) Elizabeth buys a coat at 25% discount from the original price of £60. How much does Elizabeth pay for the coat after the discount?

(3) What is 30% of £700?

(4) Calculate 40% of £800

(5) Find 15% of $400

(6) Fatima buys some perfume at 30% discount. The original price is £10. How much does Fatima actually pay after the discount?

Answers to Practice Questions 7

(1) £100

(2) £45

(3) £210

(4) £320

(5) $60

(6) £7

Chapter 8   Function Machines and Formulas

A formula describes the relationship between two or more variables (things that change). Consider a simple case first.

Example 1:

In the function machine below what is the output if the input is 2

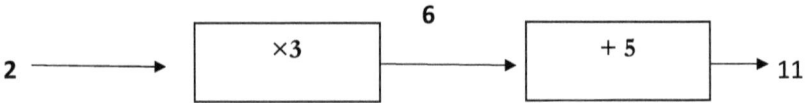

$$6$$

2 ⟶	×3	+ 5 → 11

Explanation: Input 2 then multiply by 3 to get 6, in the second box add 5 to the result to get 11.

Example 2: The function machine below calculates the steps you do in day as recorded by a pedometer. The recommended steps to walk is 10000. (a) How many steps does John do on this particular day if he walks for 85 minutes (b) Does he meet the recommended target?

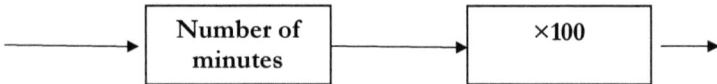

⟶ Number of minutes	×100 →

(a) **Explanation:** Input 85 in the first box then multiply by 100 in the second box to get 8500 steps as the result. So John walks 8500 steps on this particular day.

(b) Since John does not reach 10,000 steps he does not meet his target

Formulas

Example 1: A company pays 40p per mile and certain meal expenses when their sales employees visit clients. The cost of claiming mileage is calculated using the formula given, payable at 40p per mile and a fixed cost of £25 for general expenses. The formula is given by **C = M× 0.4 + 25**, where **M** represents the number of miles travelled and **C** represents the total cost in pounds payable to the employee by the company.

So for example if an employee has to travel 40 miles from her home to the client, the employee can claim 80 miles altogether for the journey to the client and back + £25 as shown below by the formula.

Using the formula we have C = 0.4× 80 + 25 = 32 + 25 = £57

(Explanation of working out: We multiply before adding. So 0.4×80 =32, finally add 32 and 25 together to get 57)

Example 2:

(1) The formula for working out the distance depends on the speed and time taken in the appropriate units.

D = S×T where D is the distance, S the speed and T is the time.

What is the distance travelled if my speed is 60kmh and I travel for 1hour and 30 minutes. 1 hour 30 minutes corresponds to 1.5 hours so, using the formula, D = 60×1.5 = 90 km.

That is, the distance equals **90km**

(2) The formula for working out the speed is given as Speed= Distance/Time

That is **S = D÷T**

Work out the average speed with which I travel, if I cover 100 miles in 2.5 hours.

Since **S = D÷T**, this means S = 100÷2.5 =40 mph (Notice the units for the first example were in kilometers and units for the second example were in miles

(3) The formula for working out time taken is given by T = D÷S

Calculate the time taken to cover 90 miles if I travel at 60mph?

Time taken, T= D÷S, so T = 90÷60 = 9÷6 =3÷ 2 = 1.5 hours or 1 hour and 30 minutes.

We have seen that formulas can be important in conversion problems

Earlier we saw the formula: S = D ÷T, that is, Speed = $\frac{Distance}{Time}$. Sometimes in the On screen questions you may be shown a distance time graph for a school coach trip and asked to work out average speed for a particular part of the journey and the time the coach was stationary. See example below

Example: A school trip by coach to a heritage site leaves at 1200 hrs from the school. The coach arrives at the destination at 1300hrs. It then stops so the pupils can look around the site. Finally after looking around the site it leaves and arrives back at school at 15:30hrs. (1) How long did the coach stop for? (2) What was the average speed on the return journey?

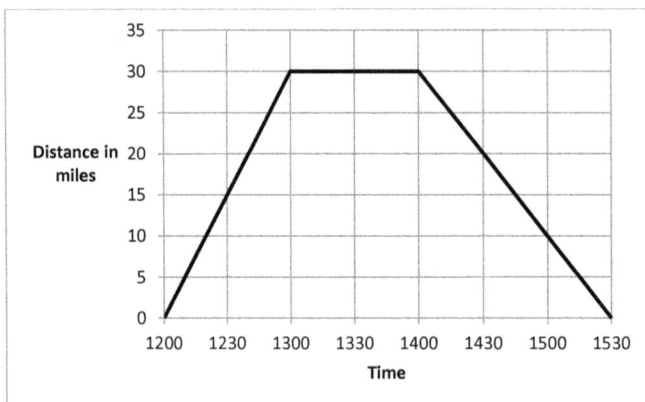

(1) From the distance-time graph above you can see it was stationary from 1300 – 1400hrs, which is 1hr

(Between these time intervals no further distance is covered, so it is stationary – see the vertical axis at 30 miles)

(2) The return journey starts at 1400hrs and ends at school at 1530hrs = 1.5 hrs.

Since **Speed =** $\frac{Distance}{Time}$, this means speed = 30÷1.5 = 20 mph

Practice Questions 8

(1) You are given that Speed $= \frac{\text{Distance}}{\text{Time}}$. Calculate my speed in m.p.h if I travel 30 miles in half an hour.

(2) In the function machine below what is the output if the input is 3

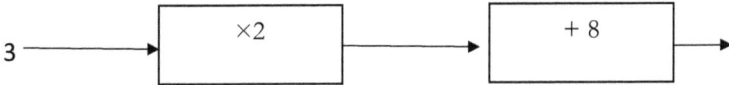

3 ──────→ | ×2 | ──────→ | + 8 | ───→

(3) Whilst she is in France Miranda wants to convert the number of kilometres she has covered into miles. She notes the formula for converting kilometres into miles is given by M = $\frac{5}{8}$X K. Assuming she covers 80 km how many miles is this?

Answers to Practice Questions 8

(1) 60 mph

(2) 14

(3) 50 miles

Chapter 9   Representing data in tables

Tables

Tables such as the ones shown below can be used to represent different types of data. This in turn can help us to interpret and analyze the information given. The examples shown below demonstrate this.

Example 1:

This table shows different languages being studied in a certain college by boys and girls

	German	French	Polish	English	Total
Boys	10	10	5	20	45
Girls	5	15	5	25	50
Total	15	25	10	45	95

Typical questions

(1) How many girls are in the French class?

Method: Look **down the** <u>French</u> **column** and **across the Girls row** as shown by the small arrows. You can see that **15 girls study French** in this college.

(2) What is the total number of boys who study languages?

Method: Going across the boys row and then down the total column you can see that the total number of boys who study languages is 45.

Example 2:

From the train time table shown below when would I reach Lichfield if I took the 13.25 train from Birmingham?

Birmingham	13.05	13.15	13.25	13.35
Erdington	13.15	13.25	13.35	13.45
Sutton Coldfield	13.25	13.35	13.45	13.55
Lichfield	13.45	13.55	14.05	14.15

Method: Starting from Birmingham at 13.25 I would reach Lichfield at 14.05 as shown.

Example 3: (Slightly harder example)

		Method of Transport			
		Car	Bus	Walking	Other
Type of Schools	Inner City	28%	32%	24%	16%
	Suburban	62%	18%	12%	8%

From the data above you can see that 32% of children take the bus in Inner City schools compared to 18% who take the bus in suburban schools. Similarly, 62% of pupils in suburban schools arrive by car as compared to 28% in inner city schools. You can also compare other modes of transport between the two schools.

Example 4:

The distances between various towns on a journey from Birmingham to London are given below in miles. What is the distance between Coventry and Milton Keynes?

	Birmingham	Coventry	Milton Keynes	London
Birmingham	XXXXXXXXX	25	70	120
Coventry	25	XXXXXXXX	45	95
Milton Keynes	70	45	XXXXXXXX	50
London	120	95	50	XXXXXXXXX

Method: Look at the column under Coventry and row across Milton Keynes. They meet at the bit shown. The distance is between Coventry and Milton Keynes as per the table above is 45 miles.

Practice Questions 9

(1) Michelle is going to see a friend in Leeds. Michelle lives in London near Kings Cross. She looks at the train time table shown below and decides to take the 14.10 train. How long is her train journey?

Depart	Arrive
London Kings Cross	Leeds
13.10	16.49
13.42	17.01
14.10	17.49
14.42	18.02

(2) Five classes raise money for a charity. The amount raised per <u>class</u> as well as the number of pupils in each class is shown in the table below:

Class	Number of Pupils	Amount of money raised in £ per <u>class</u>
A	27	37.50
B	22	32.75
C	28	40.20
D	21	22.50
E	24	24.80

(a) How much is raised by the pupils in class D?

(b) How much is raised by class B?

(3) A Deputy Head created the following table showing the number of pupils in each year group who had music lessons. How many pupils have music lessons in year 9?

Year Group	Number of pupils	Number of pupils who have music lessons
7	92	10
8	101	18
9	105	14
10	96	13
11	102	11

Answers to Practice Questions 9

(1) 3h 39m

(2) (a) £22.50

   (b) £32.75

(3) 14 pupils have music lessons in year 9

Chapter 10

Shapes and Spaces

Some common shapes

Triangles

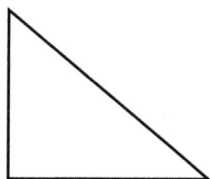

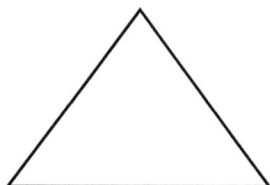

4 sided shapes

	Square	Rectangle	
All sides are equal			Opposite sides are equal

Perimeters, Areas and volumes of common shapes

Consider the shapes below:

(1) Rectangle: The shape of a rectangle is shown below:

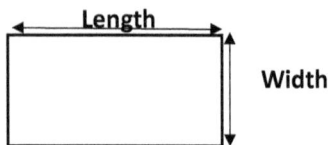

Length

Width

Perimeter of a rectangle = (distance around the outside of the rectangle) = 2×length + 2×width

Area of a rectangle = Length X Width

Note: Area is measured in units squared, e.g. cm^2 or m^2 and perimeter (distance all round a shape) is measured in the appropriate units e.g. cm or m

As explained above perimeter is simply the length around the outside of a shape measured in appropriate units. (Kilometres, metres, centimetres, millimetres, feet, inches, etc.)

Example 1: Find the perimeter of the rectangle whose length is 7metres and width is 3 metres as shown below:

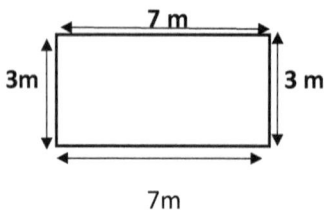

7 m

3m   3 m

7m

Method: The perimeter means the distance all the way round the rectangle. Which is 7m + 3m + 7m + 3m = 10m + 10m = 20m

Example 2: John wants to measure the perimeter of his garden. The shape and measures of the garden are shown below. What is the perimeter of the garden?

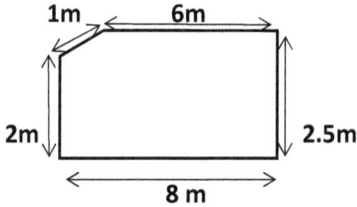

Method: To find the perimeter of the garden John simply has to add all the sides shown around the shape of the garden. So the perimeter is 1m + 6m + 2.5m + 8m + 2m = 19.5m

Areas

Area is simply the space inside a shape measured in square units.

Example 1: Find the area of the rectangle below:

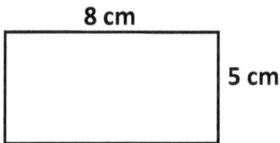

Method: Area of a rectangle is length × width. This means the area of the rectangle shown is 8×5 = 40 cms squared or 40 cm^2. (Remember area is measured in units squared.)

Triangle

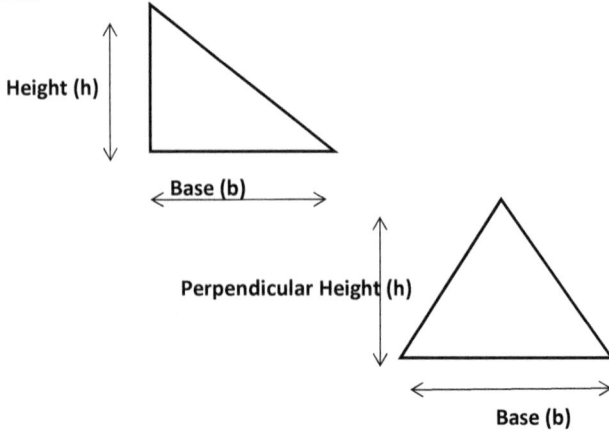

Area of a triangle =1/2 × base × height or $\frac{b \times h}{2}$ (The height is the perpendicular height relative to the base. Perpendicular means it meets the base at right angles or 90°)

Example: Find the area of the triangle shown below:

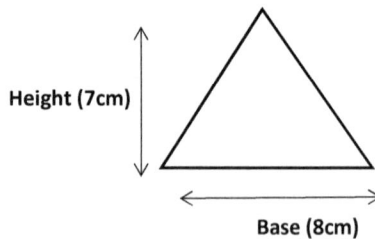

Method: Area of a triangle is given by: $\frac{1}{2}$× **base × height.**

Area = $\frac{1}{2}$ ×8×7 = **28** cm^2. So area of the triangle above is **28** cm^2

Volume of a cuboid (box shaped)

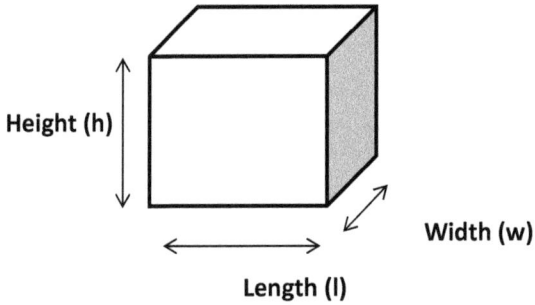

Volume of a cuboid is Height × Length × Width or V = h×l×w (units cubed e.g. cm^3 or m^3, etc)

Example: Find the volume of a box whose dimensions are shown below:

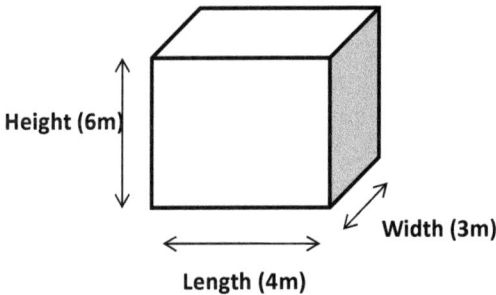

Method: Volume of a cuboid or a box as shown above = Height × Length × Width

So the volume of this box is 6×4×3 = $72m^3$

(Note: volume is measured in cubic units)

Practice questions 10

(1) Find the perimeter of the rectangle whose length is 12cm and width is 9cm as shown below.

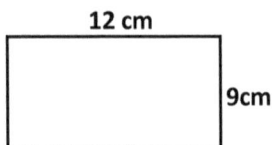

12 cm

9cm

(2) Find the area of the rectangle above

(3) Mary decides to have a flower bed in the middle of her rectangular garden shown below. The flower bed has an area of $12m^2$. The rest of the garden is lawn. What is the area of the lawn?

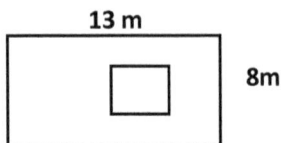

13 m

8m

(4) Peter wants to use his box for some books that he has. He works out that on average he can put 18 books in a space of $0.25m^2$. How many books can he put in the box shown below?

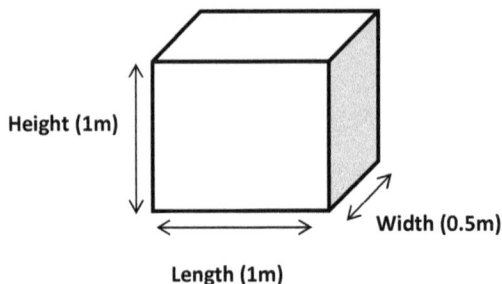

Height (1m)

Width (0.5m)

Length (1m)

Answers to Practice Questions 10

(1) 42 cm

(2) 108 cm^2

(3) 92 m^2

(4) 36 books

Common 2 –D Shapes (some of which we saw earlier.)

Square	Rectangle	Triangle

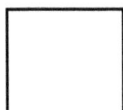 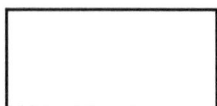 

Parallelogram	Rhombus	Trapezium

Pentagon	Hexagon	Octagon

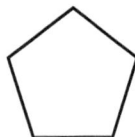 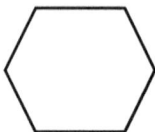 

(You can see that the pentagon has 5 sides, hexagon has six sides and an octagon has 8 sides. In this case all the sides for these polygons are equal.)

Kite **Circle** **Semi-Circle**

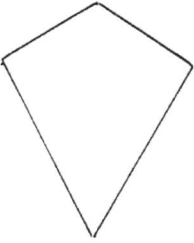 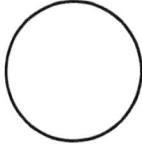 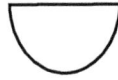

Common 3-D Shapes

Cube **Cuboid**

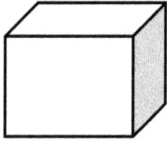 

Square Based Pyramid **Cylinder**

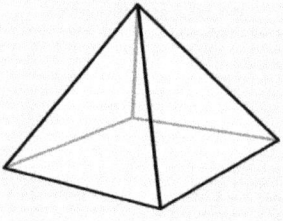 

Nets and their corresponding 3D shapes

When a **3D shape is opened out flat** you should be able to **fold it up again to get the same shape**.

Net of a cube

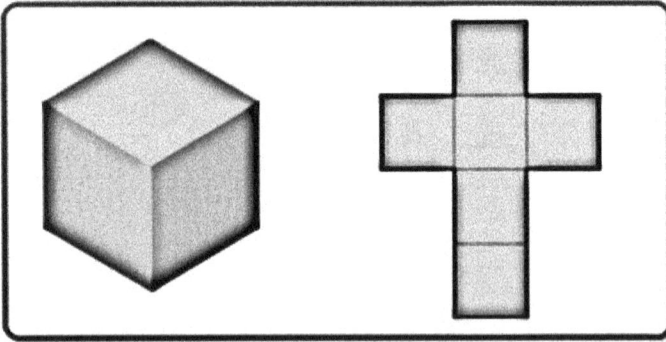

You will find that in a square shaped cube where all the sides are equal the net looks like it is shown on the right. If you fold this up you will get the solid cube again.

Some more examples are shown below:

Net of a cuboid

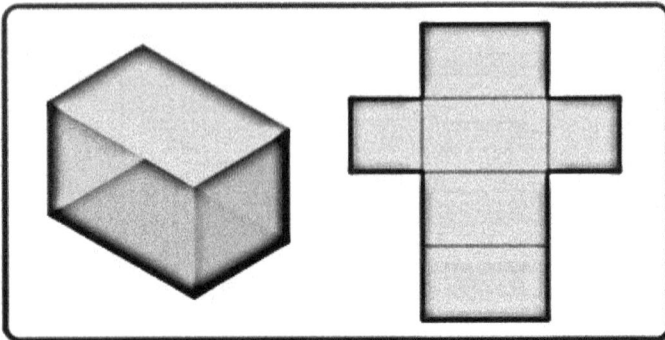

Net of a square-based pyramid

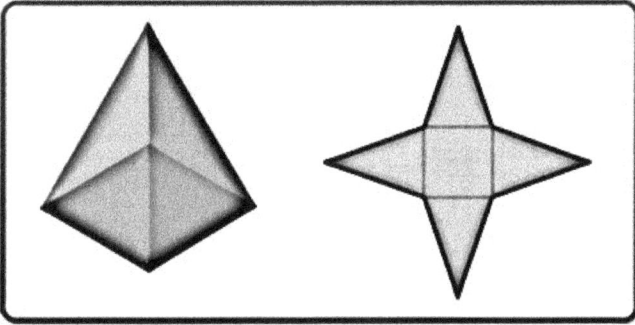

Lines of symmetry.

Some shapes have one line of symmetry others have more and some have none! If you can reflect (or flip) a half of a figure over a given line and that part of the figure remains the same, then the figure has a line symmetry. The line that you reflect over gives you a mirror image of the other half.

As we will see some shapes have more than one line of symmetry and some shapes have no line of symmetry

Examples of shapes with one line of symmetry. (The dotted line is called the line of symmetry.

Example 1:

Example 2:

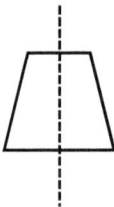

Example 3:

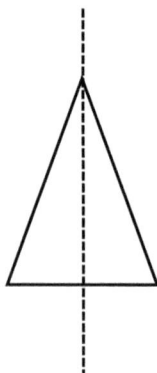

Shapes with two lines of symmetry

Example 1:

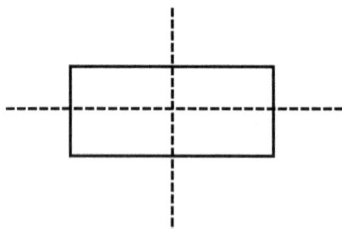

Example 2:

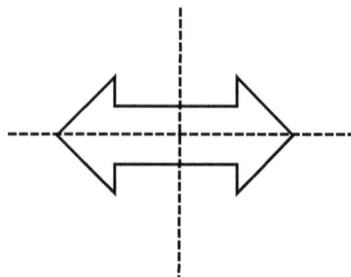

Example of a shape with no lines of symmetry.

You cannot draw a line which will reflect one half of the shape

Plans:

A plan is simply a drawing of what your room, garden or a building will look like.

A simple plan of a garden with a lawn and section for flowers is shown below.

Example 1:

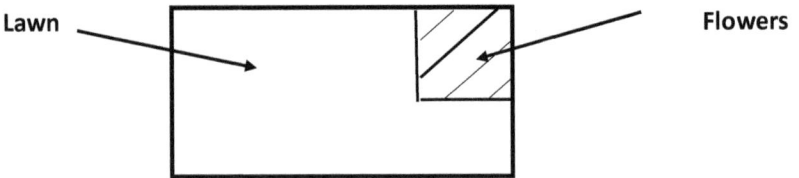

Lawn

Flowers

Example 2:

A 2D plan for example of someone's floor.

REF

KITCHEN
12'-10" x 10'-1"

BATH

BEDROOM
9'-11" x 11'-4"

LIN

REAR
ENTRY

MICRO

DW

CL

HALL

CL

CL

LIVING ROOM
15'-9" x 17'-3"

BEDROOM #1
13'-6" x 13'-10"

UP

CL

ENTRY

Example 3:

Jeremy decides a plan for his garden. He wants a small area for a circular pond an area for some flowers and a lawn. The plan is shown below. The area of the pond is $7.1m^2$. The area allocated for the flowers is $10m^2$. What is the area of the lawn?

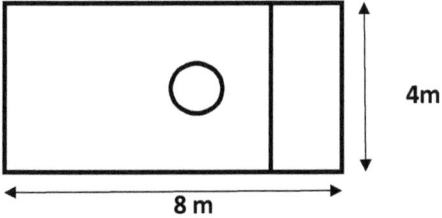

4m

8 m

Method: Area of garden = length × breadth = 8×4 = $32m^2$

Since area allocated for flowers is $10m^2$ and the area of the pond is $7.1m^2$

This means the area of the lawn is 32 – 10 – 7.1 = $14.9m^2$

Tessellations

A tessellation is a pattern made by fitting together shapes that leave no gaps.

Example 1: Tessellation with same sized triangles.

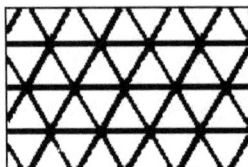

Example 2: Tessellation with same sized squares

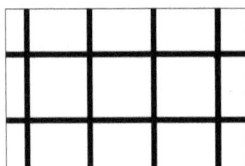

Example 3: Tessellation with a regular polygon

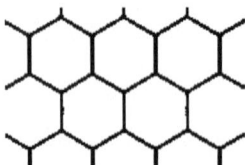

A six sided regular polygon is called a hexagon

Example 4: Tessellation with triangles and squares (Notice they join together in such a way that there are no gaps between the shapes.

Other Tessellations

You can even have **curved shapes** that join together so long as there are no gaps between the joining shapes.

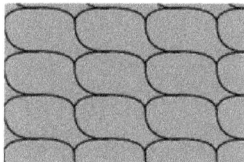

Practice questions 11

(1) What type of a 3D shape can I make from the net below?

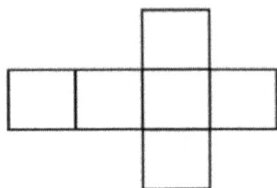

(2) Draw a line of symmetry for the 2D shape below:

(3) What is meant by a tessellation?

Answers to Practice Questions 11

(1) A cube such as the one shown:

(Note all sides are of equal length)

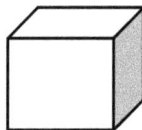

(2) The dotted line of symmetry is shown below:

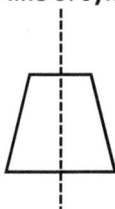

(3) A tessellation is a pattern made by fitting together shapes that leave no gaps

Chapter 12   Data Interpretation

Mean & Range

Mean: Is the sum of the numbers in a data set divided by the number of values in the Data Set

Range: This is the difference between the highest and the smallest numbers in a data set

Example 1:
Find the mean value of the following data set:
2, 7, 1, 1, 7, 8, 9

Method: Find the sum first
2 + 7 + 1 + 1 + 7 + 8 + 9 = 35
Now divide this total by 7, since this is the total number of numbers
So, 35/7 = 5
Hence, the mean value of this data set is 5

Example 2:
Find the Range of the data set 3, 5, 7, 1, 8, and 11

Method: Find the difference between the biggest and smallest numbers
So the Range = 11 – 1 = 10

Example 3:
Seniors in a care home are encouraged to walk around a bit more every day. They are each given a pedometer which records the number of steps taken. The number of steps achieved by a group of 7 seniors is recorded as shown below:
5000, 1500, 2000, 1400, 1400, 3000 and 4100 steps

What is the range in steps?

The range is simply the difference between the **highest** and **lowest** steps. So the range is 5000 – 1400 = 3600 steps

Pie Charts

When data is represented in a circle this is called a pie chart. Basically you need to remember that a full circle or 360 degrees represents all the data (or 100% of the data). Half a circle or 180 degrees represents half the data (or 50% of the data), and similarly 25% of the data is represented by 90 degrees or a quarter of a circle. Essentially, each sector or slice of the pie chart shows the proportion of the total data in that category.

Example 1:

The pie chart below shows the percentage of applicants who got different grades in a psychometric aptitude test when applying for a job in a particular company. The requirement to be short listed for a second interview was to pass with high marks. If 140 applicants took this test how many of them were short listed?

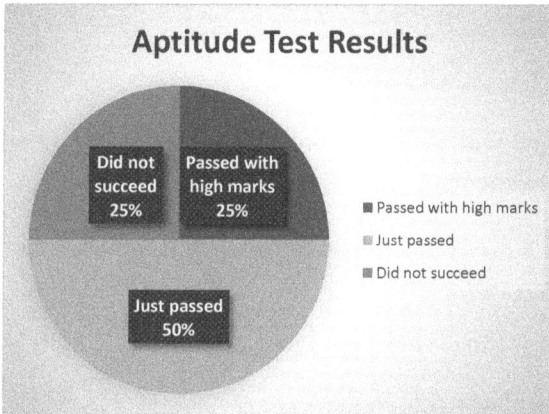

Aptitude Test Results

Did not succeed 25%
Passed with high marks 25%
Just passed 50%

■ Passed with high marks
▨ Just passed
■ Did not succeed

Method: As illustrated the results in this aptitude test for this particular company show that 25% got the required 'high marks' to be short listed for a second interview. Since a quarter of a circle corresponds to 25%. This means a quarter of the 140 applicants attained this which corresponds to 35 people.

Example 2:

The destination of 120 pupils who leave year 11 in School B in 2012 is represented in the pie chart below. The numbers outside the sectors represent the number of pupils

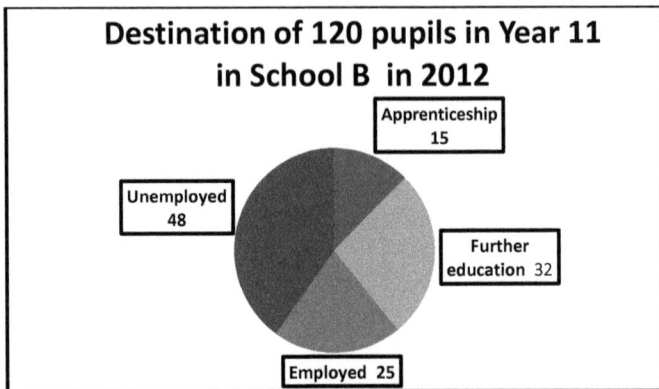

Destination of 120 pupils in Year 11 in School B in 2012

Apprenticeship 15

Unemployed 48

Further education 32

Employed 25

(1) What is the percentage of pupils who are unemployed?

Method: The number of pupils out of 120 that are unemployed is 48. So the percentage of pupils who are unemployed is $\frac{48}{120} \times 100 = 40\%$

(**Reminder**: Put 48 in a calculator, divide by 120 and then multiply the answer by 100)

(2) What fraction of pupils go on to Further Education?

Method: The fraction of pupils that go on to further education is $\frac{32}{120} = \frac{16}{60}$ $= \frac{8}{30} = \frac{4}{15}$ the fraction representing this in its simplest form is $\frac{4}{15}$ (Keep dividing the top and bottom of $\frac{32}{120}$ by 2)

(3) What percentage of pupils is either employed or in apprenticeships? Give your answer to one decimal place?

Method: Total number of pupils who are either in employment or apprenticeships = 25+15 =40, hence the percentage is $\frac{40}{120} \times 100 = 33.3\%$

(**Reminder:** Put 40 in a calculator, divide by 120 and then multiply the answer by 100)

Bar charts

Bar charts can be represented in columns or as horizontal bars. They can be either simple bar charts that show frequencies associated with data values or they can be multiple bar charts to allow for comparisons between data sets as shown below. The examples below illustrate some of the ways bar charts can be used to represent data.

Example 1: In a cosmetics shop the number of items that were sold for four top brands over a one month period were recorded as shown in the bar chart below.

(1) Which brand had the highest sales? **You can see from the column bar chart below that Brand D had the highest sales as 40 items of this brand were sold during one month, which is higher than any other brand**

(2) What was the proportion of sales for Brand D compared to the total? Give your answer as a fraction in its lowest terms. **The number of Brand A items sold were 20, Brand B were 35 and Brand C were 25 and as we saw earlier 40 items of Brand D were sold. This means the total number of cosmetic items sold during this one month period = 120. Since 40 items belonged to Brand D, compared to the total this is $\frac{40}{120}$ which simplifies to $\frac{1}{3}$**

Number of cosmetic items sold by Brand over a one month period

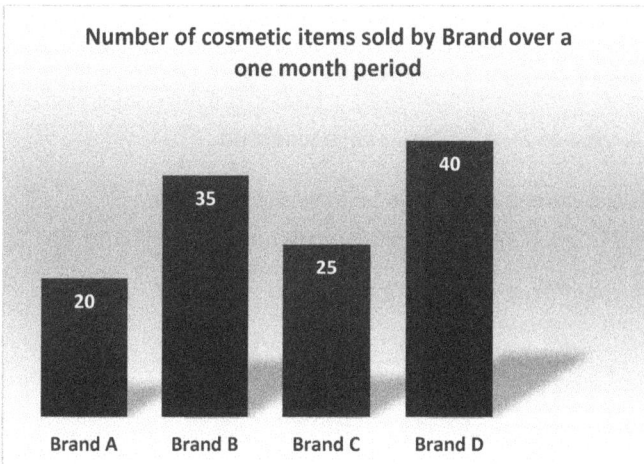

			40
	35		
		25	
20			
Brand A	Brand B	Brand C	Brand D

Example 2:

The bar chart below shows the amount of time in hours John, Bob and Bill spend surfing the web at weekends. What is the mean time per boy that is spent surfing the web at the weekend?

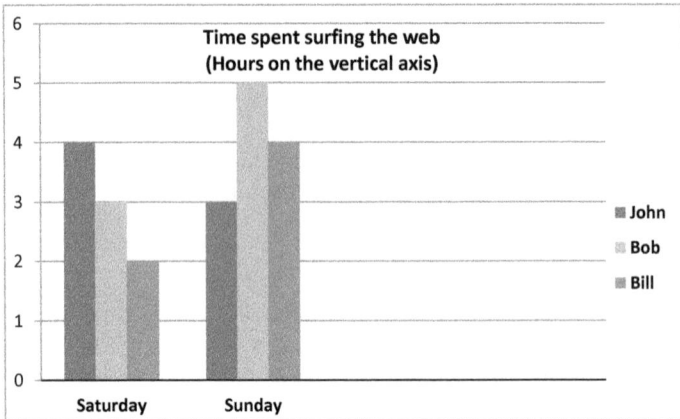

Method: John spends 4 hours on Saturday and 3 hours on a Sunday: a total of 7 hours

Bob spends a total of 3 hours on Saturday and 5 hours on Sunday: a total of 8 hours

Similarly, Bill spends a total of 2 + 4 = 6 hours on a weekend

Total time spent surfing between the 3 boys on a week end is 7+ 8 + 6 =21hours

Hence the mean time spent per boy is 21 ÷ 3 =7 hours

Practice questions 12

(1) What is mean of 3, 7, 8, 2, 9 and 1?

(2) Find the range of the data set 2, 6, 9, 1, 4, and 15

(3) In the Pie chart below what fraction of students did not succeed?

Maths Test

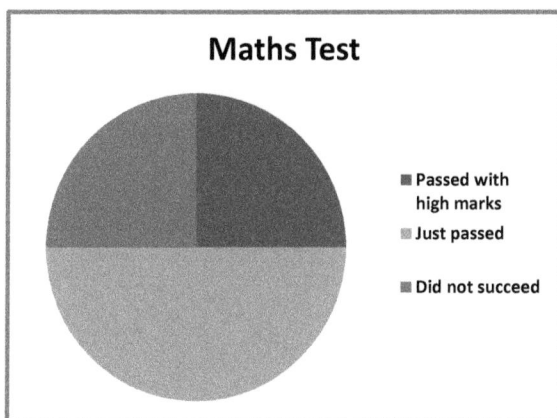

- Passed with high marks
- Just passed
- Did not succeed

(4) The bar chart below shows. How many employees are there in Company D?

Number of employees in four small companies is shown vertically

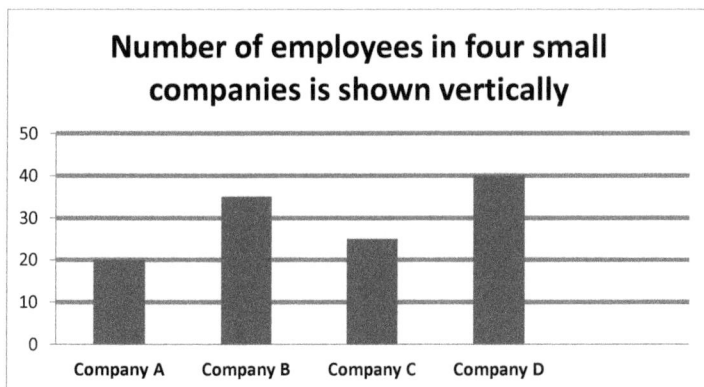

| | Company A | Company B | Company C | Company D |

Answers to Practice Questions 12

    (1) 5

    (2) 14

    (3) $\frac{1}{4}$ or 0.25

    (4) 40

Chapter 13   Probability

Probability is the likelihood or chance of an event happening.

Example1: It is very likely to rain at least one day in April.

Example 2: There is a 50% chance that if a fair coin is thrown it will land heads

Example 3: There is no chance of picking a red ball from a bag containing 6 blue balls

Example 4: It is certain that you will pick up a red ball from a bag containing all red balls.

Mathematically: **Probability is defined as the likelihood of an event happening. Probability lies between 0 and 1.**

A probability of 0 means that an event will <u>definitely not happen</u> or <u>it is impossible to happen</u>. Likewise a probability of <u>1 means it is certain to happen</u>. Probability is usually expressed as a fraction, a decimal or a percentage.

Consider two simple cases: There are 4 blue balls in a bag. You take out a ball at random. (1) What is the probability that the ball you pick is red? (2) What is the probability that the ball you pick is blue? Although this is a trivial example you can see that in question (1) it is impossible to pick red ball since all the 4 balls in the bag are blue. Hence the probability of picking up a red ball is 0. Similarly, in question (2) the probability that you pick a blue ball is 1. That is you are certain to pick a blue ball, since all the four balls are blue.

Many events of course happen with a probability between 0 & 1. For example a probability of 0.9 would indicate a high chance of an event happening, whereas a probability of 0.1 would imply a low probability of an event happening. The probability of an event happening is defined as:

$$\frac{number\ of\ ways\ in\ which\ the\ event\ can\ happen}{total\ number\ of\ outcomes}$$

Example 1:

Summary:

(1) The probability of an event happening lies between 0 and 1.
(2) Probabilities can be expressed as a fraction, decimal or a percentage
(3) The probability of an event can never exceed 1.
(4) A probability of 0.1 or $\frac{1}{10}$ or 10% means it is not so likely to happen
(5) A probability of 0.9 or $\frac{9}{10}$ or 90% means it is very likely to happen
(6) A probability of 0.5 or $\frac{5}{10}$ or 50% means there is an even chance of the event happening

Practice questions 13

(1) Chloe meets up with Anne who is usually a few minutes late. Is Anne likely to be (a) on time (b) early (c) a bit late

(2) The weather forecast says that the chance it will rain some day in April is almost certain. Is it likely to not rain in April or rain in April?

(3) 1 in ten million people win the national lottery every week. Fatima buys a lottery ticket. (a) Is she certain to win (b) certain to lose (c) very unlikely to win.

Answers to Practice Questions 13

(1) Probably a bit late

(2) Likely to rain

(3) Very <u>unlikely</u> to win

Exam Type Questions

HOLIDAYS

(1) Joanna and Jane decide to go on holiday to Barcelona from Manchester. They book their flights online. The costs are shown below:

Cost of return flight to Barcelona from Manchester: £45 per person
*************** Online discount: £15 per person ****************

   (a) How much does Jane pay for her ticket if she books online?

   (b) How much do Jane and Joanna pay in total if they book online?

(2) The cost of going on a Mediterranean cruise for 7 days is going for £475 per person listed on icelolly.com. There is a £115 discount per person if they book straight away. Peter and his wife decide to go on this break as they feel it will do them good and it seems affordable, so they book straight away. How much does the cruise cost in total?

(3) John likes taking city breaks. This time he decides to go to Rome for 3 days which includes half board and return flight. Because he books online he gets a 10% discount. Also he changes £110 to Euros and gets and exchange rate of 1.2 Euros to a pound.

Cost of return flight (half board 3* Hotel) to Rome from London Heathrow: £140 per person
***************Online discount: 10% per person ****************

   (a) How much does John pay for this city break?

   (b) How many Euros does John get for his £110?

STARTER BUSINESS

(4) Peter decides to print and distribute 3000 leaflets for his painting and decorating business. He charges £110 per day. He gets 12 days of work from this advertising. His printing and distribution of leaflets costs £150.

   (a)   How much in total will he earn before his printing and distribution costs?

   (b)   How much does Peter earn after his printing and distribution costs?

MEASUREMENTS

(5) Paul does some tiling in his Kitchen. The wall he tiles is 3m wide and 4m long. He uses square tiles in which the lengths of the sides are 30cm by 30cm. How many tiles will he need to get the wall tiled? (Give your answer in whole number of tiles)

(6) Brian is visiting a friend in Wakefield. Brian lives in Birmingham. He looks at the map and measures the possible routes to get to Wakefield. His measurements on the map show that his best option comes to 30 cm on the map. He makes a note of the scale which is 1cm for every 4 miles.

(a) How far in miles is Wakefield from Birmingham.

(b) Because of the traffic he averages 40mph to get to Wakefield from Birmingham, how long does the journey take?

(7) Peter belongs to a walking group that walks 16 Km every week. If 8km is approximately equal to 5 miles, estimate how many miles the weekly walk consists of?

SALES & BUSINESS

(8) The table below shows the total sales by a fashion retailer in their shops in London, Manchester and Leeds in Millions of pounds.

Total Sales	2010	2011	2012	2013
London shops (In Millions of £)	15	14	16	17
Manchester shops (In Millions of pounds)	7	8	8.4	8
Leeds shops (In Millions of pounds)	8	9	8.5	8.7

(a)   What was the total sales in 2011 for all three cities?
(b)   What were the sales in Manchester in 2012?
(c)   What was the increase in sales in Leeds from 2012 to 2013?
(d)   Was Manchester or Leeds doing better in 2011?

(9) A company calculated that it had given bonuses to its junior and senior staff in the ratio of 1:3. There was a total of £4000 bonus given to all the staff. Assuming there were 20 junior staff, how much did each member of the junior staff get?

(10) There are 20 employees in a small company. Two of them go on a special training course. What is the fraction of employees that go on this training course? Give your answer as a fraction in its simplest form.

MISCELLANEOUS QUESTIONS

(11) Joanna is raising funds for a Cancer Charity. She manages to persuade 12 people to give £3.50 each for this charity. What is the total amount she collects?

(12) A group activity consists of 15 tasks. Each task lasts 10 minutes. How many hours will this group activity last?

(13) Fatima has organized a meeting that begins at 10:00. She has planned to start with a general introduction for 5 minutes, a power-point presentation for 10 minutes and finally a question and answer session for 25 minutes. When does the meeting end? Give your answer using the 24-hour clock.

(14) In a certain apprenticeship scheme, the probability of an individual being accepted is 80%. Is a person likely, certain or unlikely to be accepted?

(15) Jamie and Susan decide a plan for their garden. They want a small circular pond and an area for some flowers and a lawn. The plan is shown below. The area of the pond is $12m^2$. The area allocated for the flowers is $15m^2$. What is the area of the lawn?

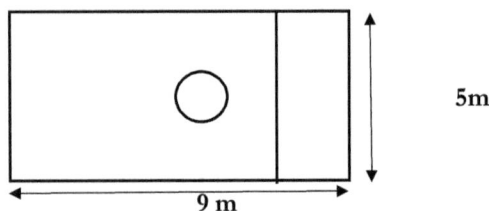

(16) Natalia earns £22,000 in the transport department working for a local authority. The new tax legislation allows her a personal allowance of £11000 before tax is deducted. On this salary her taxable income is 20%. How much tax does Cathy pay?

(17) The bar chart below shows the percentage of employees who earn £20,000 per year in Company A. What was the mean percentage of employees who earned £20,000 from 2008 to 2011?

Percentage of employees who earned £20,000 per annum in Company A

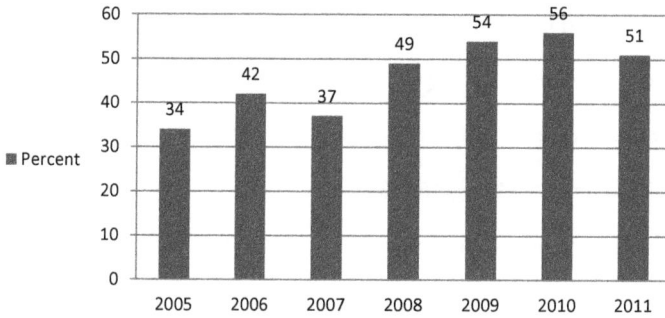

(18) Mariam produces hand knitted baby jumpers at home which she then sells on ebay. The costs and selling price of the three types of baby jumpers she makes are given below. What is her profit from the blue jumper she makes?

	Cost of Making	Selling Price on ebay
Red Jumper	£2.60	£6.00
Blue Jumper	£3.80	£6.27
Green Jumper	£2.90	£5.50

Answers to Exam Type Questions

(1) (a) Answer: Jane pays £30

Method: Simply subtract £15 from £45 so 45 – 15 =30

(b) Answer: They pay a total £60

Method: Total cost without discount = 2×45 = £90

Total discount = £15 + £15 = 30

So the Total Joanna & Jane pay = 90 – 30 = £60

(2) Answer: Total cost of holiday is £720

Method: Total cost = 2×475 - 115×2 = 950 – 230 = £720

(3) Answer: (a) The city break costs £126

Method: Cost of holiday – 10% discount = 140 – 14 = £126

(b) John gets 132 Euros for his £110

Method: 110×1.2 = 132 Euros

(4) Answer: (a) £1,320

Method: (a) He charges £110 per day. So for 12 days work should get 12×110 = £1320 before costs

Answer: (b) £1,170

Method: Earnings – costs = £1320 - £150 = £1,170

(5) (b) Answer: 134 tiles

Method: Total area of wall to be tiled is 3m×4m, since each metre = 100cm

area of wall = 300×400 = 120000cm^2. However, each tile has an area of 30×30 =900 cm^2. So the number of tiles required is 120000÷900 =133.33 tiles! But he can't buy fractions of tiles so Paul needs to buy 134 tiles.

(6) Answer: (a) 120 miles

Method: Since 1cm =4miles, this means 30cms = 30×4 = 120miles

Answer (b) 3 hours

Method: Distance travelled is 120 miles, average speed is 40mph so time taken = 120÷40 = 3hours

(7) Answer: 10 miles

Method: Since 8km is 5 miles, then 16km is 2×5 = 10 miles

(8)(a)   Answer: £31M or £31,000,000

Method: Total Sales in 2011 in all three cities were: 14 + 8 + 9 = £31M

(b) Answer: £8.4M or £8,400,000

Method: In the sales table look under 2011 vertically and Manchester horizontally, they cross at £8.4M

(c) Answer: £0.2m or £200,0000

Method: Sales in Leeds in 2012 is £8.5m, in 2013 it is £8.7M so the increase is 8.7 − 8.5 = £0.2M

(d) Leeds

(9) Answer: £50

Method: Total parts in ratio 1: 3 is 4. This means each part = £1000. Since there are 20 junior staff, so each junior staff member gets 1000÷20 =£50

(10) Answer: $\frac{1}{10}$

Method: $\frac{2}{20}$ go on a training course. This simplifies to $\frac{1}{10}$

(11) Answer: £42

Method: 3.5×12 = £42

(12) Answer: $2\frac{1}{2}$ hours or 2.5 hours

Method: Total time = 15×10 = 150 minutes =$\frac{150}{60}$ hours = 2.5 hours

(13) Answer: 10:40

Method: Total time taken is 5 + 10 + 25 = 40 minutes

So the meeting ends at 10:40

(14) Answer: Likely to be accepted

Method: 80% chance is quite high so the probability or chance that an individual is accepted for an apprenticeship is likely.

(15) Answer: $18m^2$

Method: Area of garden = 5×9 =$45m^2$

Area of pond = $12m^2$, Area taken by flowers = $15m^2$.

Hence total area taken by pond and flowers =12 + 15 = $27m^2$.

So area remaining for the lawn is 45 − 27 = $18m^2$

(16) Answer: £2,200

Method: She pays tax on (22,000 − 11,000) = £11,000

Since the tax payable is rated at 20% this means Natalia pays 20% of

£11000 = $\frac{20}{100}$×11000 = £2,200

(17) Answer: 52.5%

Method: Add up the percentage of employees who earned £20,000 between 2008 & 2015 and divide by 4. That is (49 + 54 + 56 + 51)÷4 = 210÷4 = 52.5%

(18) Answer: £2.47

Method: Blue jumper costs £3.80, Mariam sells it for £6.27

Profit = 6.27 − 3.80 = £2.47

Some Useful Reminders and getting ready for Functional Maths 2

Natural Numbers: are {1, 2, 3, 4, ...}

Whole Numbers: are {0, 1, 2, 3, ...}

Integers: These are whole numbers that include both positive and negative numbers including 0. So for example ....-5,-4,-3,-2, 0, 1, 2, 3, 4, ... are all integers.

Multiples: These are simply numbers in the multiplication tables.

For example the multiples of 6 are 6, 12, 18, 24, 30, ...

Factors: A factor is a number that divides exactly into another number as for example, the number 2 in the case of even numbers.

3 is a factor of 9, as 3 goes exactly into 9. Other factors of 9 are 1 and 9.

15, has two factors other than 15 and 1. The two factors are 5 and 3, since both these numbers go exactly into 15. **Example:** Find all the factors of 21. The factors are: 1, 3, 7 and 21 (since all these numbers divide exactly into 21)

Prime numbers: A prime number is a natural number that can be divided only by itself and by 1 (without a remainder). For example, 11 can be divided only by 1 and by 11. Prime numbers are whole numbers greater than 1. So for example the first 10 prime numbers are: 2, 3, 5, 7, 11, 13, 17, 19, 23 and 29. **Be careful that an odd number is not necessarily a prime number.** For example **9 is not a prime number** as its factors are 1, 3 and 9 and **prime numbers should have only two factors, 1 and the number itself. Also, note that 2 is a prime number, the only even number that can be divided by 1 and itself!**

Lowest Common Multiple (LCM)
This is essentially the smallest number that will divide exactly by the numbers given. Consider the examples below:

Example 1: Find the LCM of 15 and 45

One method is to find the multiples of both numbers and identify the lowest common multiple as shown below:

Multiples of 15 = 15, 30, **45**, 60, 90, ...

Multiples of 45 = **45**, 90, 135, 180, ...

Clearly **45** (the highlighted number above) is the smallest number that is divisible by 15 and 45.

Example 2: Find the LCM of 10 & 15

First find the multiples of each number:

Multiples of 10 = 10, 20, **30**, 40, 50, 60, 70,.....

Multiples of 15 = 15, **30,** 45, 60, ...

You can see that **30** is the lowest common multiple since it is divisible both by 10 & 15.

Highest Common Factor (HCF)

This is the biggest number that will divide exactly into all the numbers given

Example 1: Find the HCF of 15 & 45

Method: Find the factors of each number given and then identify the biggest number that will divide into both these numbers as shown below:

Factors of 15 ={1, 3, 5, **15**}, Factors of 45 ={1, 3, 5, 9, **15**, 45}

You can see that **15** is the **highest common factor** which divides into **both** 15 and 45 exactly.

Example 2: Find the HCF of 8 and 32.

First find the factors of each number given

Factors of 8 ={1, 2, 4, **8**}, Factors of 32 = {1, 2, 4, **8**, 16, 32}

You can see that the number 8 is the highest **common** factor which divides into 8 and 32 exactly.

Square numbers and square roots

Squaring a number is simply multiplying a number by itself.

So 4^2 means $4 \times 4 = 16$, 12^2 means $12 \times 12 = 144$ and so on.

The square root is written like this $\sqrt{}$ and means finding a number which when multiplied by itself gives you the number inside the square root.

Example1: Find $\sqrt{16}$. The answer is clearly 4. Since 4×4 =16

Let us consider some other square roots.

$\sqrt{49}$ = 7, $\sqrt{121}$ =11, $\sqrt{100}$ =10, $\sqrt{225}$ = 15,

$\sqrt{256}$ = 16, $\sqrt{324}$ =18

Cubes

Cubing a number is simply multiplying the number by itself three consecutive times. A cube of a number is written as x^3, where x is the number.

So, for example, 5^3 means 5×5×5 =25 × 5 =125

Similarly, 6^3 =6 × 6 × 6 = 216, 7^3 =7 × 7 × 7 = 343, 9^3 = 9 × 9 × 9 = 729,

10^3 = 10 × 10 × 10 = 1000

Cube Roots

Cube roots are found by finding a number which when cubed gives you the number inside the cube root.

So for example the cube root of 125 is written as $\sqrt[3]{125}$

Also we know that 5X5X5 =125, so that $\sqrt[3]{125}$ = 5

CPSIA information can be obtained
at www.ICGtesting.com
Printed in the USA
LVHW01s2301150917
548880LV00017B/373/P

Functional Skills Maths Level 1

By Vali Nasser

!G

Copyright © 2016

E-book editions may also be available for th s title. For more information email: valinasser@gmail.com

ISBN-13: 978-1539534754

ISBN-10: 1539534758

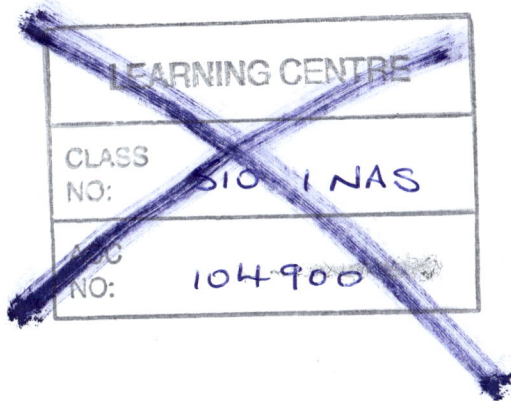

Table of Contents